WASHI TAPE

WASHI TAPE

101+ IDEAS FOR PAPER CRAFTS, BOOK ARTS, FASHION, DECORATING, ENTERTAINING, AND PARTY FUN!

Courtney Cerruti

Quarry Books
100 Cummings Center, Suite 406L
Beverly, MA 01915

quarrybooks.com • craftside.typepad.com

For Matt

© 2014 by Quarry Books

First published in the United States of America in 2014 by
Quarry Books, a member of
Quayside Publishing Group
100 Cummings Center
Suite 406-L
Beverly, Massachusetts 01915-6101
Telephone: (978) 282-9590
Fax: (978) 283-2742
www.quarrybooks.com
Visit www.Craftside.Typepad.com for a behind-the-scenes peek at our crafty world!

10 9 8 7 6 5 4 3 2 1

ISBN: 978-1-59253-914-7

Digital edition published in 2014
eISBN: 978-1-61058-037-4

Library of Congress Cataloging-in-Publication Data

Cerruti, Courtney.
  Washi tape : 101+ ideas for paper crafts, book arts, fashion, decorating, entertaining,
and party fun! / Courtney Cerruti.
      pages cm
  ISBN 978-1-59253-914-7 -- ISBN 978-1-61058-037-4 (eISBN)
1. Tape craft. 2. Gummed paper tape. 3. Masking tape. 4. Japanese paper. I. Title.
  TT869.7.C47 2014
  745.59--dc23

                                                                    2013038949

Design: Rita Sowins / Sowins Design
Photography: Matthieu Brajot

Printed in China

Contents

INTRODUCTION › WHAT IS WASHI TAPE? ... 9

› WORKING WITH WASHI TAPE ... 10

› WASHI TAPE STORAGE ... 14

WASHI TAPE PROJECTS

› ON PAPER ... 17

› ON GLASS & CERAMICS ... 33

› AS DÉCOR ... 41

› AS FASHION ... 55

› FOR CELEBRATIONS ... 59

› FOR KIDS ... 71

› FOR EVERY DAY ... 93

› IN THE OFFICE ... 107

› SEASONAL ... 113

RESOURCES ... 124

ACKNOWLEDGMENTS ... 126

ABOUT THE AUTHOR ... 127

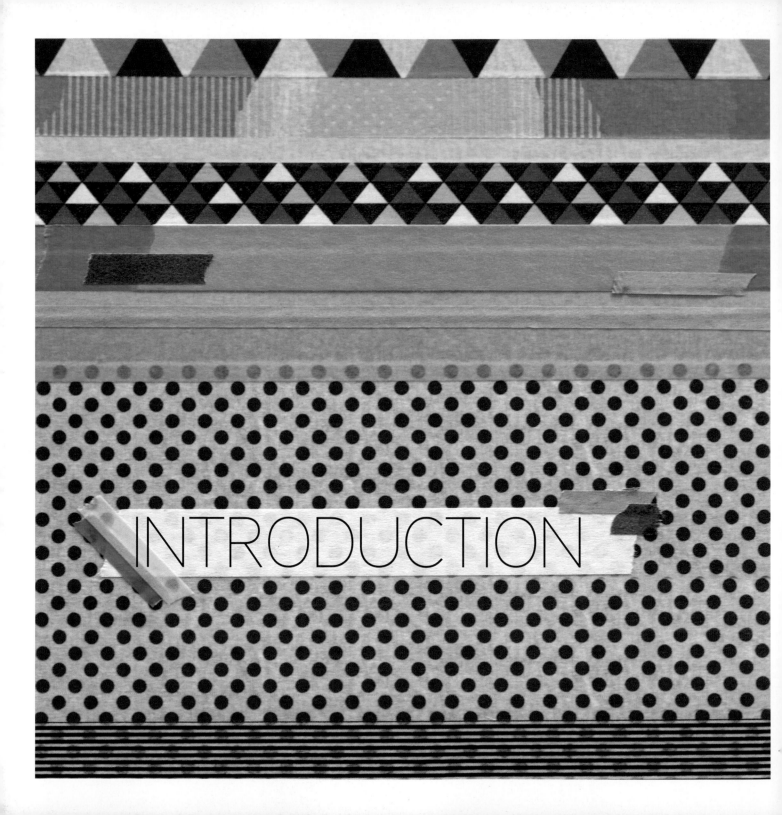

INTRODUCTION

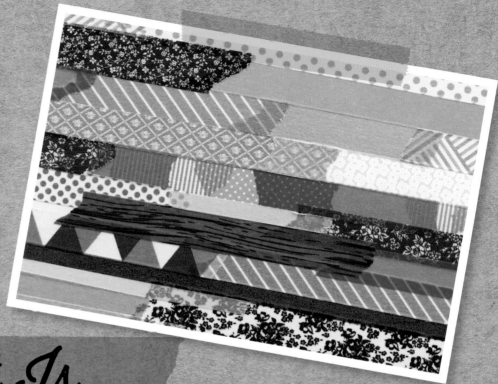

What Is Washi Tape?

Washi tape is a low-tack, decorative paper tape that can (and should) be used on everything! The word *washi* refers to the Japanese paper from which the tape is made. Known simply as masking tape in Japan, this decorative tape is slightly transparent and removable, making it a popular choice for all types of art and craft applications. While washi tape originated in Japan, it is now made throughout the world. Changes in style and quality come with this variety of manufacturers. The majority of washi tape rolls have either a solid color or a pattern, measuring ½″ (l cm) wide, with a satin finish.

You can also find tape in thinner and thicker widths, in metallic and neon styles, as well as with glossier finishes. MT, a leading washi tape manufacturer, recently launched a

line of tape called MT Casa that includes rolls up to 12" (30 cm) wide. Other brands also offer decorative tape that they call washi, but it may not have the same properties. In addition to rolls, you can find it in sheets or prepunched shapes. Buy what you're attracted to, play with it, and see if it works for your project; above all, PLAY!

Working with Washi Tape

Washi tape can be torn or cut. It can be layered for various effects, as most of it is slightly transparent. It can be used to create large designs for home décor or placed on a quick piece of correspondence for an instant pop of color. I love to tear washi tape. The deckled, irregular torn edge can add to the handmade look of a project. You can also cut the edges of your tape with scissors for a clean edge, use pinking shears for a zigzag edge, or tear the tape with a tape dispenser, which will give you a tiny serrated edge. To create shapes like small triangles, squares, or arrow-tipped ends use scissors to freehand cut your tape.

When making detailed shapes or trimming tape in place (for example, on a surface), use a dull craft knife to score the tape, and then tear it away for a clean line. You can punch washi tape with craft punches to create little shapes that can be used like stickers.

You can work with washi tape on waxed paper to freehand cut or punch. To create your own patterned sheets, layer tape onto waxed paper, overlapping strips of tape slightly as you go. Create a solid sheet of all one kind of tape or vary colors, widths, and patterns to create your own design. Freehand cut shapes with scissors, cut detailed shapes with a craft knife and cutting mat, or punch shapes with craft punches.

There is no wrong or right way to use washi tape!

Punch shapes from purchased sheets of washi tape.

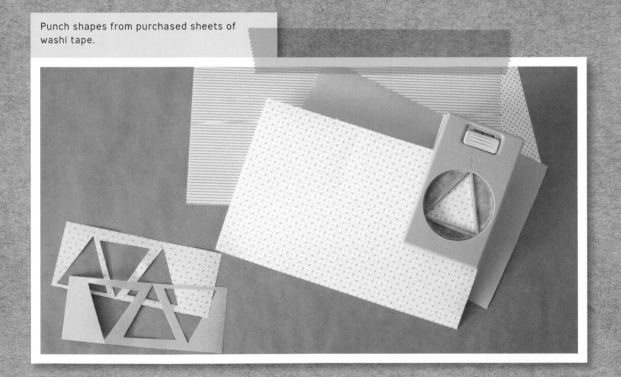

Make your own patterned tape sheet.

Punch shapes on handmade sheets of tape.

Washi tape can be used on almost anything. It is ideal for paper and is removable on most types, with the exception of some kraft papers, textured fibrous papers, or vintage papers. You can use the tape on glass, ceramic, metal, and other surfaces. Always test a tiny area of whatever you plan to decorate to ensure that the tape will stick and not damage the surface. Washi tape works best on flat surfaces or 3-D objects with straight sides. If using on antique or vintage papers, know that most washi tape will pull away the fibers of brittle and old papers. Do a little test before you lay on the tape.

Because washi tape is low tack, it is not ideal for permanent use on items exposed to extremely hot or cold temperatures, as the tape will most likely peel off after a while.

Many washi tapes are slightly transparent and can have different effects, depending on the surface you apply them to. If you're using a pale or light tape on a dark surface, consider doubling up the tape for added opacity and saturation or placing a strip of white tape down first, and then topping it with the light-colored tape. (See photo at right.)

WRITING ON WASHI TAPE

Washi tape has a lightly waxy surface that will only take permanent inks. I prefer to use ultrafine Sharpies or 0.5 Micron pens when adding messages or text. Allow the ink to dry a few moments before placing the tape on your project. Black and brown inks show up best, but experiment with various ink colors for different effects.

Washi tape is readily available and inexpensive, allowing everyone the chance to play with design and color, change their environment, and get creative with little time or investment. Washi tape is adaptable and fun, making it the perfect ad hoc design element for everyday use.

Layering washi tape colors and patterns can yield a variety of results.

Washi Tape Storage

Once you start using washi tape you might find yourself surrounded by piles and piles of brightly colored rolls. You'll want to curate your pretty new addiction in an organized fashion so you can grab a roll easily for your next project. Here are just a few ideas to help you corral your collection.

A thread spool organizer doubles as the perfect storage for a large collection of washi tape. You can find these at the local craft store, and they can hang on the wall or stand alone on a desk. Organize your tape by color or at random!

For easy and stackable storage, cigar boxes are a clever way to house your tape. I love collecting boxes of varying sizes and searching out one with interesting images and bold graphics.

These lightweight picture ledge shelves from IKEA are a sleek and inexpensive way to customize your craft area. Stacks of washi tape stay organized and add a pop of color to any room.

Transform a box for waxed paper or aluminum foil into a container for your washi tape. The blade at the lip of the box will allow you to cut tape easily and keep rolls tidy. I use these all the time for workshops and craft nights!

To create a storage container, punch a hole at either end of the box. Add a ¼" (6 mm) dowel that is 1" to 2" (2.5 to 5 cm) longer than the length of the box. You can sharpen the ends of the dowel with a pencil sharpener for a more finished look. Add your rolls of tape before placing the dowel into the box. You can easily make a more permanent version of this container by adding a fine edged saw blade to the edge of a wooden box that fits the rolls of washi tape.

WASHI TAPE

PROJECTS

On Paper

Adding washi tape to paper is a quick and satisfying way to create a pop of color or pattern on any item. Use the tape to personalize everyday office and stationery products. Add a burst of neon pink to a neutral card or a swash of metallic gold to a black envelope. Washi tape and paper make a perfect pair!

Look for tapes with special sentiments like "hello," "thank you," or "miss you."

PATTERNED POSTCARD

Turn a piece of 4" x 6" (10 x 15 cm) card stock into a quick and stylish postcard. Use washi tape to create your own design. Vary thick and thin tapes and alternate between patterns and solids to create a dynamic composition.

Add characters or little critters by using tape with animals or stylized creatures printed on it. Add a stamp and pop it into the mail!

BIRTHDAY CARD

Whip up a cute little cupcake for a birthday card. Using 1½" (3.8 cm) -wide solid tape, cut out a frosting shape and layer a patterned cupcake wrapper made from patchwork tape on top. Complete with a washi tape candle. Make an enclosure card with the same methods in miniature.

A single cupcake makes a statement, while three little cupcakes are as sweet as can be.

Use rubber stamps to create your own special greeting.

CELEBRATION CARD

Buntings are the epitome of party décor. Make a card that says "celebration" instantly. Draw three softly swooping lines across your card. Cut triangles from various solid and patterned washi tapes to create bunting flags of all styles.

Play with color and pattern to create a bunting card for any occasion.

Place the flags along your drawn lines to create three-tiered buntings. The tape is low-tack, so you can move the flags around until you have the perfect arrangement.

WELCOME HOME

Make a card for a housewarming gift or for someone returning from a long trip or a year away at school. Creating little houses is easy with squares and triangles cut from a length of tape. Add a strip of grass for extra color. Stamp or write your message!

Make a village of little houses in a winter scene. You can send them for holiday greetings or just because.

PATCHWORK

Use washi tape to lay out a design. Working with the tape in a sketchbook is a great way to test color and shape. Use patterned and solid tape that represents fabric colors to create a patchwork design for a quilt. Use the same method of playing with color and pattern for fashion, paper crafts, or pattern making.

Working with washi tape is easy when designing. Because the tape is low tack, you can move it around the page as your design develops.

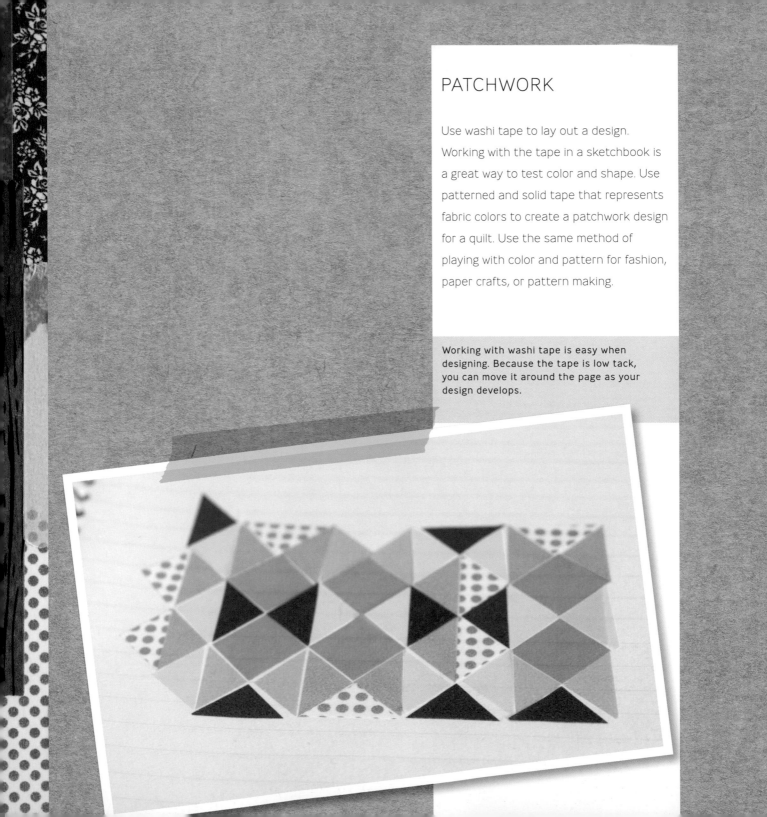

Use this technique on other paper goods to create gift tags or food packaging.

PERFECT PLAID

Plaid is one of my favorite patterns. Using washi tape to create your own plaids is a great way to start playing with color without pressure. You can let go and experiment with the transparency of the tape and learn how solids create unique colors and patterns as they overlap. There are no rules for making fun and funky plaids. Create cards, postcards, or little framable works of art.

QUICK AND EASY THANK YOU

I always keep a couple of tiny thank-you notes in my purse; you never know when someone will surprise you and you'll want to say thank you for being awesome. A little square of paper goes from scrappy to special with a border of washi tape. Tuck a thank-you note into a pocket, leave it on a seat, or add it to a bouquet of wildflowers.

Pair tiny thank-you cards with tiny envelopes. Seal with washi tape and give freely!

Use bits and bobs from your washi tape stash to add color to your present presentation. Even the littlest scrap can be used to make these tags into gift toppers for boxes, bags, and stockings.

FOR YOU!—GIFT TAGS

Use mailing tags from the office supply store to make quick and easy gift tags.

SKETCHBOOK COVER

Personalize a moleskin notebook and make it your own. A blank journal can easily be turned into an instant personal treasure when washi tape creates a unique design or theme. You can also customize a cover for the perfect gift. Play with various types of tape to create covers for every occasion, including travel, memories, and special events.

Solid tapes create bolder graphics, while delicate patterns like florals or muted tones will give a softer, lighter feel.

Using washi tape to add color and pattern is an ideal way to jump-start your creative process when flipping to a blank page.

MEMORIES AND MEMENTOS—ART JOURNALING

Keeping an art journal or daily record of your life can feel like a lot of pressure on some days. Using washi tape for both decoration and utility is a surefire way to attack a couple of pages before dinner. Add little bits of paper ephemera like bus tickets, movie stubs, photos, and notes with washi tape. Use tape to frame out quotes or thoughts, edge pages, or hold down photos.

Cut a half circle in pink or red to create sweet rosy cheeks for your photo (A).

Block out parts of your photo with color or pattern. Lay strips of tape over the area you want to block out, overlapping strips slightly as you go (B). Trace the shape, in this case the collar, lightly with a pen or pencil (C).

Carefully remove the tape from the photo and place on waxed paper. Cut along the traced line to create a sticker mask for your photo shape. Place the tape sticker back onto the photo (D).

COLORIZE VINTAGE PHOTOS

Transform vintage or found black-and-white photos into modern works of art. Use washi tape to enhance or block out various portions of an image and add color.

Because washi tape is low-tack, you can place overlapping strips of it on waxed paper, and then cut them into a shape. With this technique, I created a patterned collar for my vintage lovely.

This technique works great for adding party hats, crowns, cheeks, or rosy lips to vintage black-and-white photos.

BIRDIE MOBILE

Mobiles can feature a gathered collection of treasured items, adding color to a space and swinging lazily in the breeze. This sweet, little birdie mobile is fun to put together and can be made in your favorite colors to match any décor!

Precut shapes like stars, hearts, letters, and animals are easy to find in craft stores. Transform these blanks into colorful patterned pieces you can use for all kinds of décor.

To create feathers, place overlapping strips of washi tape on card stock in roughly the size you want your feather to be. Freehand cut a feather shape. Flip over the feather and apply more washi strips to the back so you have a double-sided feather. Trim off any excess tape.

Repeat these steps to create your bird shape.

To create the texture in the feather, cut into the shape from the outside edge at an angle toward the center, creating a fringe to mimic the spines of a feather.

Use a stick, your shapes, a bit of string, and any other elements like beads or pom-poms. Tie your mobile elements on the stick at varying lengths. Add ribbon or string to hang, and then place in a sunny spot to enjoy!

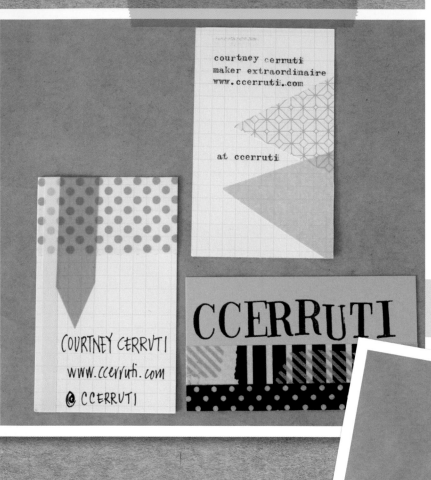

CALLING CARDS

Before hashtags and handles, people gave out calling cards at parties, gatherings, and social events. Make on-the-fly calling cards for your next craft night, cocktail hour, or impromptu date. Punch shapes to create a graphic or use strips of tape to decorate blank business cards.

Use stamps or a typewriter or handwrite the necessaries for your modern calling card.

On Glass & Ceramics

Sometimes you need that last-minute touch to bring a project all together. Because washi tape is removable on glass and ceramic surfaces, you can coordinate china, customize flatware, and add the last detail without having to buy matching everything. Here are some ideas for making your party décor or breakfast in bed extra special.

COFFEE MUG

Perfect for breakfast in bed, washi tape turns a plain, glass coffee mug into a sweet morning surprise. I love the little kitties saying "Hello" as the first greeting of the day. Pair a delicious mug of coffee or tea with a tasty baked good for the ideal breakfast combination.

Quickly add a special message or sentiment to any beverage glass with a scrap of tape.

Switch out art easily and often. Consider framing vintage ephemera, photos, or even postcards. Add washi tape to accent your framed collections.

FRAME

A classic white frame is clean, neutral, and the perfect canvas for a quick and easy décor switch. Change the look of your framed art with a washi tape border. For these botanical prints, I added warm-toned, gridded tape that gives a little nod to the scientific nature of this art.

Look for tape with printed messages for special occasions or select floral, lacy tape for a romantic feel.

VOTIVES

Keep clear glass votives on hand for events or small occasions or even to create instant atmosphere at dinner. Transform the look of basic votives with a few strips of washi tape. Patterned, metallic, or even graphic tape (like the road map) sets your table alight with a wash of color and warmth.

Combining tape with simple glass vases will tie décor together and create an eye-catching centerpiece quickly and easily.

BUD VASES

A trio of vases to hold freshly picked blooms makes a statement with a washi tape design. Add a punch of color that complements your florals, making them a focal point.

CERAMIC SERVING PLATTER

White dishes make the perfect clean palette for a pop of color or a shimmer of gold. Add strips of washi tape to plates and serving platters at every event. Make an offering of macaroons even more special with a stripe of mint green, pink, and gold tape studded with stars.

Crisscross tape to create an asymmetrical design. Use odd numbers of colors and stripes for more impact.

Who says white dishes are boring? Change your look for every event by playing with color and pattern.

SWEET SUCCULENTS

Turn a simple terra-cotta pot into a stylish gift on the go. Perfect for a housewarming or to use as a favor, your favorite potted plant (succulents are lovely!) gets a graphic punch from washi tape.

Use small potted plants as place cards on the table. Place a strip of solid tape on each pot and write the guests' names.

As Décor

Want to try out a pattern or bright color at home? Because washi tape is removable, you can fearlessly add it to walls, furniture, fixtures, and more without the commitment of paint! Brighten a room with a splash of color or add a giant wall graphic with tape as ad hoc art. Modernize your grandparents' dining room chairs with a neon stripe. The possibilities are endless!

PHOTO WALL

Curate your favorite daily snapshots quickly and easily into a clean, graphic presentation. Using washi tape to hang each small print makes it a snap to arrange and rearrange these little moments into perfect seasonal collections.

It's easy to let tear sheets and photo inspiration pile up in a corner. Create a rotating inspiration wall by hanging all your collected ephemera with washi tape.

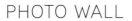

LIGHT 'EM UP!—SWITCH PLATE COVER

Love those kitschy and colorful switch plates . . . but not forever. Add tape when the fancy strikes, and then remove it when your taste changes. Use stripes, polka dots, metallic, or neon for fun and fanciful fixtures.

A washi tape switch plate is a clever way to add color to a room. Ideal for rented spaces, the tape can be removed when you need a change of décor.

INSTANT ART

Blank canvas or wooden blocks can be so intimidating. Overcome your fear and just start taping! Scraps of washi tape create abstract art for any room. Layer solids and patterns to create work you are proud to hang. Consider juxtaposing cut edges with torn ones and balancing out verticals and horizontals.

Cover the edge too. It's easiest to line up your tape with the front edge of the canvas and work your way around the entire edge. Allow excess to go over to the back side and miter your corners as you go for a clean finish.

For permanent art pieces that might go in sunlight or a steamy space like a bathroom or kitchen, seal in the tape by brushing on a matte medium or spraying it with an acrylic varnish.

PRIVACY WINDOW

A soft and elegant pattern can provide color and privacy. Cut or punch shapes into patterns for windows or doors. Hand-cut teardrop shapes make lovely vines along the base of this window, but turn these upside down and they could be falling raindrops for a taller window. Play with scale and shape to make unique decals for decorating a window or giving privacy.

Make washi tape patterns on glass kitchen cabinets or closet doors to hide the clutter.

Outside looking in: Because the tape is transparent, light will still shine through your window, and color will be visible from both inside and out.

How wonderful would it be to wake up to a birthday message, a greeting of "I love you," or even a happy face to start the day?

Great for kids, a washi tape graphic makes getting ready in the morning fun!

SECRET MESSAGE

Surprise someone with an unexpected accent. Add a heart made of tape to window blinds for a sweet and loving surprise.

Simple shapes or words on boring blinds create a sweet surprise. This is lovely indoors but can also be used to keep the neighbors guessing!

BATHROOM MIRROR

Waking up to a sweet surprise like a heart on the bathroom mirror sets the tone for your whole day. Add a star for a birthday surprise or a crown to start the day as a queen or king.

TILES

Add a pop of color to any tiled surface, like a backsplash or bathroom wall. Layer strips of washi tape along a tile here or there for an unexpected punch. Use a dull craft knife to trim tape just to the edge of the tile along the grout. Areas with extreme heat or exposure to moisture and water will require more frequent tape changes, which means you get to change your décor whenever the mood strikes!

Adding tape to existing tile is an excellent way to test out a new color before investing in a remodel. Place strips on tiled or painted walls to get an idea of color before you commit to a change.

VIGNETTE WALL

Make quick and decorative frames with washi tape to feature small bits of ephemera and works of art. Create a simple square around a photo and then miter the corners with scissors to soften the look of the frame. Create a patterned background by laying down strips of tape on waxed paper. Determine the orientation of your artwork to the pattern you created with the tape. Trim the tape sheet into a diamond, circle, or irregular organic shape. Tack small photos onto the wall with simple triangles cut from tape. Play with color in your vignette.

Create a tape background. Trim the background to the desired shape (A).

Cut and prepare to place on the wall (B).

Peel away the waxed paper and place directly on the wall. Cut washi tape into various shapes and sizes for interesting framing options (C).

Write your own message with a solid strip of washi tape and a Micron or Sharpie pen.

Color-code your tape labels to easily find types of books, like arts and crafts, reference, and literature.

COVERED MATCHBOX

Candles are a sweet and lovely impromptu gift for a friend, a hostess, or yourself on a cozy night at home. Don't ruin the look of a beautiful candle with a liquor-store box of matches. With a few strips of tape, make the box just as lovely!

Layer several strips of softly patterned or solid tape to cover the matchbox graphic. Add a small swatch of tape with a lovely message, a pretty image, or a sweet sentiment for the finishing touch. Remember not to cover the strike strips!

BOOK ORGANIZATION

Corral a cluttered shelf by wrapping books in solid paper. Kraft paper makes the perfect neutral, but you can wrap books in patterned papers or brightly colored solids for a child's room. Add labels to the spines with washi tape and a Micron pen.

KRAFT LETTERS

These are available at most craft stores, and they are perfect for personalization. In a collagelike manner, add tiny bits of washi tape over your monogram for a patchwork letter you can place on a bookshelf, hang on the wall, or use for party décor. A brightly colored monogram looks sweet in a child's room, a set of initials makes a perfect pair for a wedding, and a single letter is great for a baby shower.

Make a letter for each family member to place on a mantel or bookshelf for a personalized piece of art.

Make coasters for any occasion. Give a set as a hostess gift when attending a party.

COASTERS

Washi tape's satin finish repels moisture. Cover blank coasters with strips of tape for fun and functional coasters for a special event or for every day. As with any paper coasters, they can easily be replaced after lots of use!

COLOR-BLOCK CHAIR

A few colors add punch to a wooden chair. Create a special seat for the birthday girl or bride and groom or just brighten up the chair you picked up at the last garage sale. Turn those hand-me-downs into modern pieces for your space.

Cut washi tape into shapes to place on the back or seat of a chair.

WALL BORDER

When I get restless with my surroundings, my mind jumps to papering my walls with vintage wallpaper or painting them with a bold color. Test your desire for outrageous design by dipping your toe in the water first. Use wide strips of tape to cut graphics you can add to the wall without worry. Try triangles, squares, or circles for a graphic punch or soft, organic shapes for a fluid design. You can add tape shapes directly to the wall and trim them with a dull craft knife on the wall (be sure to score only the tape and not cut your wall). Or you can place large strips of tape on waxed paper and cut on a cutting mat or with scissors to create your wall shapes.

Because this tape is low tack, if you don't like the placement of your design, you can peel it off and move it until it's in the right spot.

Use this method to create a faux headboard or focal art piece for a room.

WALL GRAPHIC

Create a large-scale graphic directly on the wall with washi tape. Cut strips in advance or cut and work as you go, composing a shape directly on the wall. Pair dark, solid colors with brights and metallics. Use the MT Casa line to create large, solid shapes like triangles or create your own sheets of tape by laying down strips on a sheet of waxed paper, and then cutting to the desired shape. (See page II.)

LAMPSHADE

Changing your décor is easy with washi tape. Add a few strips of tape to a basic shade to punch up the design. Start from the top of the shade, moving down, rather than across for a clean design.

LAMP BASE

Coordinate your shade with a solid base. Run tape along the shape of the base, allowing its curves to dictate the design.

Washi tape allows you to personalize basic lamps on a budget.

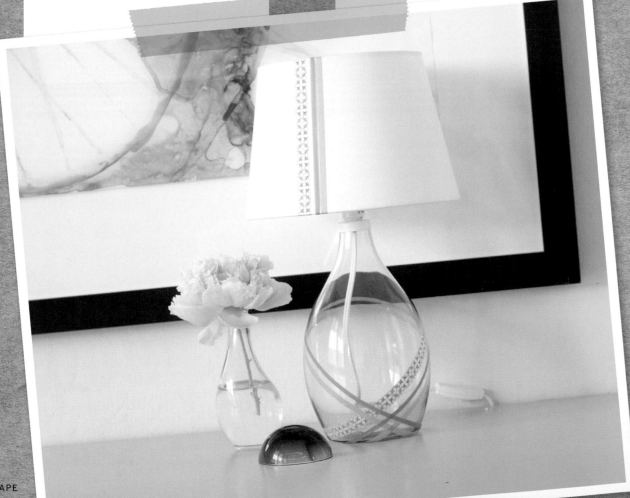

As Fashion

Fashion trends change almost as quickly as you tear the tags off that newly purchased outfit. Change your accessories as often as your mood with washi tape. Add color and pattern to basic blacks and solids.

Try placing tape on the toes of your shoes too!

FOOT FORWARD

When you need a little glam for the night, add metallic washi tape to the heel of your shoes. A triangle or strip of tape gives you that missing detail. This is a great way to coordinate your shoes to your outfit without matching all your accessories.

If you have bangles with curved sides, you can wrap your tape around the bracelet like you would wrap string on a bobbin, moving the tape from inside the bracelet to the outside. Wrap at a 45-degree angle as you move along the bangle.

To trim away tape in more intricate shapes, use a craft knife for precise cutting.

BOLD BANGLES

Make that perfect accessory when you're getting dressed: Add tape to a wooden bangle. Wrap a rounded bracelet or neatly place a strip on a bracelet with straight sides . . . pastels for day wear and gold for a night out. Change your jewelry as often as your outfit!

GEOMETRIC EARRINGS

Blank wooden jewelry is a great foundation for a washi tape design. Geometric shapes like circles, triangles, or diamonds can be layered with strips of tape for eye-catching patterns. Silhouettes of animals and objects can also be great to transform for the day with a layer of tape.

NAIL ART

Cut a small graphic from washi tape and use it on your next manicure. Place a heart or other shape onto the last layer of colored polish just as it becomes tacky. Add a clear topcoat to lock in your design. You can also use tape to create a mask. Use a strip on dry nails to mask off an area of your nail, paint over it, and then remove it to reveal a clean strip or tip.

Experiment with solids, neons, and patterns for various effects.

For Celebrations

The perfect event is made memorable through the details. Use washi tape to add pattern and color to cutlery, create an impromptu table runner, or make a wall graphic for the birthday girl or happy couple. Coordinating last-minute design details is easy and fun with washi tape.

CELEBRATION BUNTINGS

Buntings are a sweet and cheerful addition to any space. Hang over a bed, window, or desk or place on a shelf to add a touch of whimsy. To create a traditional bunting with single-point flags, fold strips of washi tape over a string or ribbon, doubling the tape back onto itself. Place flags of various patterns and colors along the string, and then cut each flag into a pointed triangle shape and hang.

My heart always melts when I see a bunting; it makes me instantly happy.

To create a bunting with double-point flags, add strips of tape in the same manner as the single-point flag bunting, and then cut a triangle away from the center of each flag, creating a double point.

Make a larger bunting of various shapes by mixing small single- and double-point flags with flags made from large strips of washi tape. Cut the large flags into scallops and other shapes. Layer patterns, colors, neons, and solids or add a message in tape with preprinted letters.

Because of the small scale of buntings, this a great project for the end of the roll when you only have a little tape left. You can also create large-scale buntings by placing washi tape on paper to create flags.

Who says your pets can't have a party? Instead of throwing an animal-themed party for kids, use washi tape to make up treat bags or decorate party hats for a pet party.

There's no need to place tea lights in containers if you add washi tape!

PAPER PERFECT

Paper cups and straws in primary colors are the perfect canvas for classic patterns and graphics. Using tape with dogs and cats (or any theme), create a suite of tableware for your next party. Add little animal flags to straws for that special detail and also as a way to identify drinks.

PARTY TEA LIGHTS

Good lighting can take an event from average to atmospheric. Use washi tape to customize inexpensive tea lights, giving your table or outdoor décor that extra attention to detail.

Let kids decorate their own invites or make them together for a family craft night.

CAMPFIRE INVITE

An impromptu summer sleepover is the perfect excuse to eat s'mores! Create a camping-themed invite to inspire a night of stargazing and roasting marshmallows over the fire.

CAKE TOPPER BUNTING

A miniature bunting looks oh-so-sweet atop a pretty cake. Using various colors and styles of washi tape, make a bunting with baker's twine or string tied to wooden skewers. Create multiples for various cakes so you have a collection for larger parties or events.

Use cake toppers year after year for a birthday tradition or create a new one every year for a birthday surprise!

CUPCAKE TOPPERS

Make cute cupcake toppers in a snap with a toothpick and washi tape. Use regular tape to create single- or double-point flags and skinny tape for miniature flags. Create coordinating sets for any occasion.

Add these little flags to pies, cookies, and other baked goods.

MOCK TABLE RUNNER

Create a table runner or bright graphic for a dinner party. Turn a gray linen into a bold statement with a geometric neon design. Use tape to add lines, chevrons, diamonds, triangles, or other sharp shapes. Place on linens or tabletops to create mock table runners or place mats.

Use this technique with paper. I love using kraft paper to cover a table, and then adding a washi tape graphic for a custom look.

PRETTY LITTLE THINGS

Who doesn't love a beautifully wrapped package? Solid kraft and black boxes are the base for this geometric design. With metallic and neon tape, these simple little gift boxes become almost too pretty to open.

Washi tape makes any gift extra special. Use it to enhance the presentation on a plain box or even on a wrapped gift.

This would be *awesome* on the hood of a new car, on the back window for a birthday or wedding, or even on the door of a new house!

SURPRISE!

Sometimes those extra-special gifts are the hardest to wrap. Because this tape is removable, you can put it on delicate surfaces without damaging them. A bow made of light and dark tape creates a 3-D effect.

CONFETTI POUCHES

Giant paper confetti turns any gathering
into a party. Tuck handfuls of colorful
confetti into little paper pouches bursting
to be opened.

Instead of confetti, make little seed packets
or glitter pouches.

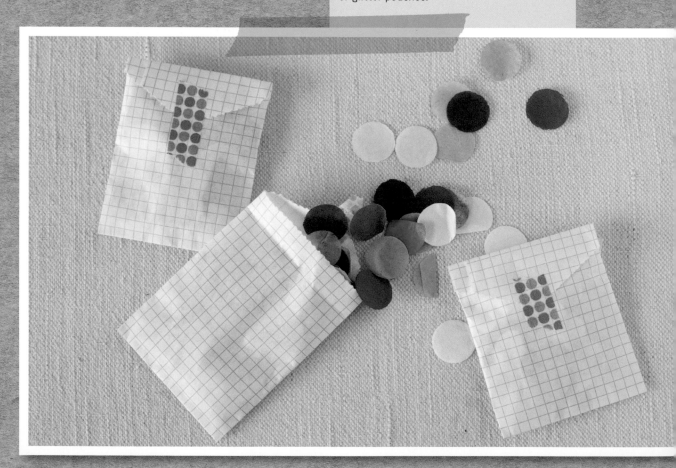

CHEESE PLATE

Cute little flags tell guests what you're serving. Use toothpicks to create markers for cheese or small bites. Use multiple flags to call out ingredients that are common allergens.

You can also place washi tape right on serving plates or bowls to label ingredients in your dishes at a potluck or to make sure your fancy serving dish comes home with you.

TABLE NUMBERS

Create 3-D–looking numbers for table markers or birthdays. Use two colors of washi tape to create your number and shadow, and then place thinner tape along all the edges for a number that pops.

Use metalics for weddings or brights for birthdays.

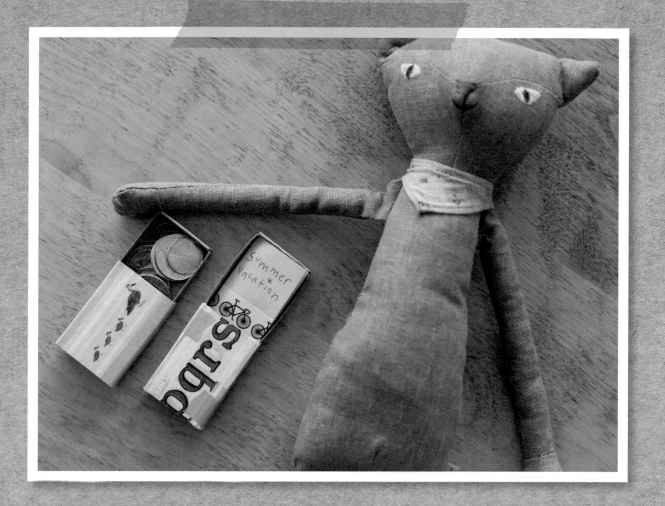

For Kids

Washi tape is a fun and easy way to get kids creating art and crafts. With small bits of tape, make cards, ornaments, frames for artwork, and refrigerator monsters! Play with color and pattern without breaking out the paint.

SANDWICH WRAP

Classic is timeless for a reason. Wrap a PB&J in waxed paper and tape it closed with patterned tape. No time for a paper wrap? Add washi tape to a plastic sandwich bag or little snack box so your little (or big) one knows you made this lunch especially for them.

Make impromptu snack bags and pouches to hold a handful of goldfish crackers or other tasty treats. Fold up a small sheet of waxed paper and seal closed on all sides with a fold and a piece of washi tape.

LUNCH BAG

I remember the first day I stayed for lunch in the cafeteria in kindergarten. I don't remember what I ate, but I do remember the note my mom tucked into my bag. Make a memorable lunch for your little ones with a washi tape decoration on a simple paper bag.

Easily mark lunches for everyone so you don't have to look twice when heading out the door.

RACE TRACK

Keep kids occupied by laying down tracks in the living room, kitchen, or hallway. With a few twists and turns, you can create a roadscape that will keep little cars and little hands moving all afternoon.

Turn up the volume on your indoor play! Create little houses from cereal boxes, cardboard, and other recyclables. Add a washi tape town to your track. You can even use the tape to create a town or scene directly on the wall. Then peel away when playtime is over.

IMPROMPTU HOPSCOTCH

Play indoors on rainy days with an impromptu game of hopscotch. Create your own little song about the rain and skip through your day.

FAIRY WAND

Make magic happen every day. Cover a wooden star shape from the craft store with strips of tape in golds, pinks, or whatever colors conjure spells for you. Wrap the tape along a wooden dowel to create the wand. Hot glue your star onto the dowel to create a magic wand for your little fairy princess.

Create a set of wands for play dates and parties. Let each child decorate his or her own wand.

PAPER CROWN

Using card stock, cover half of the length of a sheet of 8.5" x 11" (21 x 28 cm) paper with washi tape, making a regal pattern. Cut the paper in half, and then snip crown points into the paper, cutting triangles about halfway down the center of the page. Tape the ends of the paper together with washi or a little glue if the paper is very stiff. Place atop the head of your prince or princess. Make a crown for a birthday, holiday, or any day.

Create a pattern with tape.

Cut points to create a crown.

Make crowns for any occasion or just because! Make larger versions for adults and use them as photo booth props.

These little crowns would also make really cute napkin rings at a fairy-tale tea party!

Create bottles of beads, sequins, or dried flower petals. Make sets of magic potions and pixie dust for gifts or party favors.

FAIRY CROWN

Make a matching crown for teddy with a toilet paper tube. Cut the tube in half with scissors, and then cover with various strips of tape. Use a patchwork method or add tape in stripes or slanted strips. Cut crown points along the edge of the tube about halfway down, making a sweet little crown for all your playtime pals.

PIXIE DUST

Upcycle little bottles into vessels for pixie dust. Fill glass jars with fine glitter and whisper a wish before closing up. Add washi tape labels for certain kinds of magic and spells worthy of any pixie hijinks.

WALL GRAPHIC LETTER

Help little ones learn their colors and letters and create art at the same time. Make a giant monogram for a door or wall with washi tape.

Play with patterned and solid combinations for distinctive effects when making letters.

CHALKBOARD INSPIRATION STARTER

Inspire little minds with a springboard for drawing. Add a washi tape cityscape, lines for writing, or other shapes and letters to a chalkboard and let kids add their own images!

Because tape is removable, allow kids to experiment with their own shapes. Include a bucket of chalk for hours of imaginative play.

Little ones love Band-Aids. Covering them with bright and happy patterns might just make a trip to the doctor's office a little easier.

WASHI TAPE BAND-AIDS

Make boo-boos less scary with bright and colorful adhesive bandages Add tape over cloth Band-Aids and trim away excess tape to make them fun and fanciful.

MEMORY MATCHING GAME

Turn little square cards into a memory matching game. Use card stock or even cut up a cereal box to create the game tiles. Add a washi tape pattern to one side of the tile, and then turn over and add an animal, plant, or object. Create pairs of matching tiles with rolls of tape printed with jungle animals, birds, or little characters.

Make alphabet cards or flash cards in the same way for an educational game or to take on a road trip.

Make a washi tape postcard to
go with your magnets!

FRIDGE MAGNETS

Cover blank wooden shapes with washi
tape to create graphic magnets for your
fridge. Use tape with vintage type for
a retro feel or use colored tape for a
graphic punch. Use strong magnets with
superglue or epoxy to create magnets
that will hold postcards, art, sketches,
and photographs all at once!

PARTY HATS

Easily make party hats by rolling colored 8.5" x 11" (21 x 28 cm) paper into a cone shape and taping it into place. Trim off the excess at the bottom. Let little ones decorate their own hats with washi tape, glitter, and pom-poms. Add string or ribbon for secure-fitting hats.

Decorate store-bought paper hats if you don't have time to make your own.

Let kids decide what art they want to display, allowing them to change out artwork freely. Great for tiny spaces or delicate surfaces, a washi gallery can hang in a hallway or even on a closet door.

ARTWORK GALLERY

Create a rotating gallery for artwork and sketches. Line the walls with washi tape frames to display work whenever creativity strikes.

MASQUERADE

Blank masks are a dime a dozen. Make
them special with washi tape. Let kids
make their own masks for dress-up,
costume parties, and Halloween!

Make the kids' table at your next dinner party extra fun with tape graphics, blank shapes for coloring in, or even stripes of color.

SNACK TIME!

Mealtimes and cleanup can be fun with a guide to where things go. Make place mats for plates—X marks the spot for a cup, or a target for where the bowl goes.

SILHOUETTE ART

Use a photo to create a keepsake silhouette of your child. Print out a picture in profile in the size you want for the final piece. Flip the print over and cover the back of the page in washi tape. Make sure to place tape well beyond the photo. Flip the paper back over to the photo side and cut along the silhouette with scissors or a craft knife. Tape or glue the silhouette onto a solid-colored paper and frame!

Cut away the silhouette from the washi tape-patterned paper.

Now you're ready to display!

Let children create their own patterned paper for a collaborative keepsake.

CLOTHESPIN DOLLS

Transform a simple wooden clothespin into a tiny doll that will keep little ones entertained for hours. Use washi tape to create patterned clothes. Use a Micron pen or colored pencils to add a face and hair. Create a little family of friends to take in the car or tuck into a backpack.

I love making these. They make a great gift topper, ornament, or special treat in a bagged lunch!

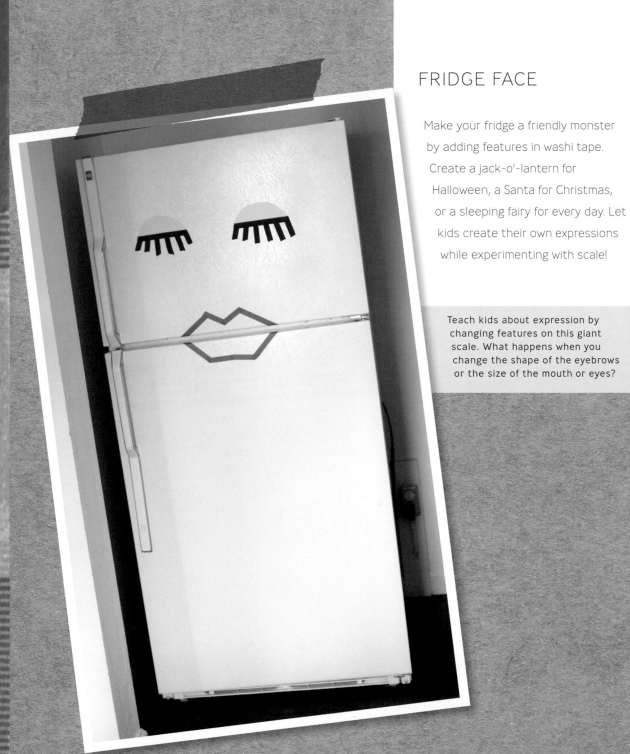

FRIDGE FACE

Make your fridge a friendly monster
by adding features in washi tape.
Create a jack-o'-lantern for
Halloween, a Santa for Christmas,
or a sleeping fairy for every day. Let
kids create their own expressions
while experimenting with scale!

Teach kids about expression by
changing features on this giant
scale. What happens when you
change the shape of the eyebrows
or the size of the mouth or eyes?

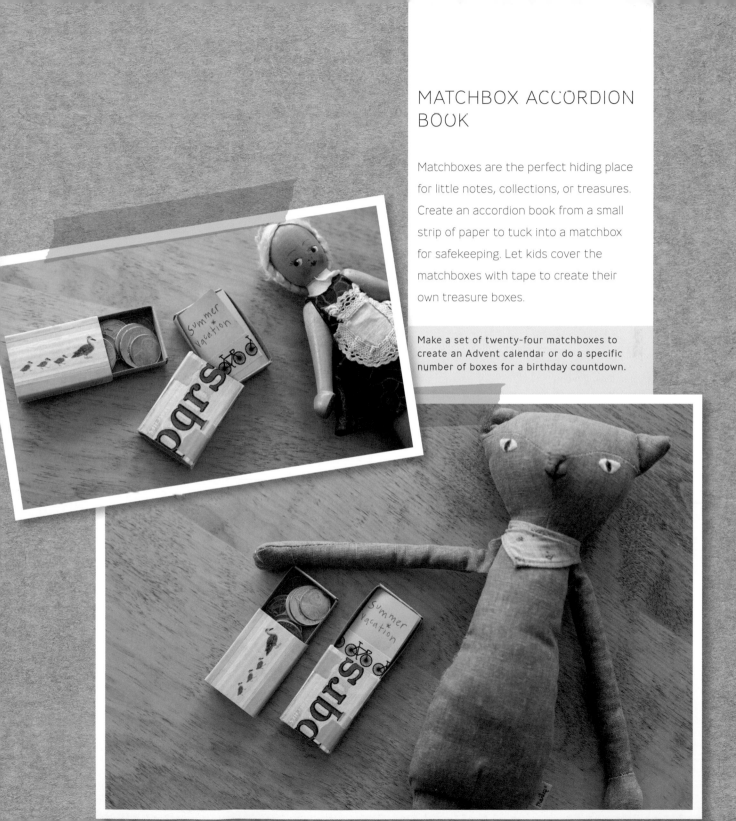

MATCHBOX ACCORDION BOOK

Matchboxes are the perfect hiding place for little notes, collections, or treasures. Create an accordion book from a small strip of paper to tuck into a matchbox for safekeeping. Let kids cover the matchboxes with tape to create their own treasure boxes.

Make a set of twenty-four matchboxes to create an Advent calendar or do a specific number of boxes for a birthday countdown.

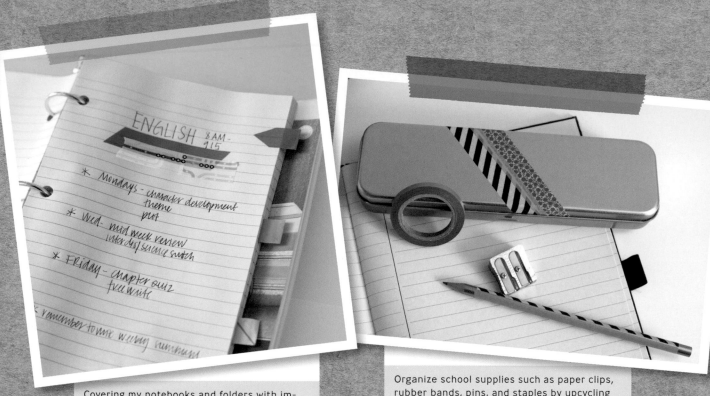

Covering my notebooks and folders with images was my favorite back-to-school activity. Personalize your notebooks, binders, or even a clipboard with tape.

Organize school supplies such as paper clips, rubber bands, pins, and staples by upcycling mint tins and labeling their contents with washi tape.

BACK-TO-SCHOOL BINDER

Color-code notes, keep tabs on assignments, and organize your information by using washi tape to mark sections in binders and notebooks.

BACK-TO-SCHOOL SUPPLIES

Make your back-to-school supplies reflect your style. Customize your pens and pencils with washi tape. Hold pens and pencils in a striking metal case decorated with tape. Add tape to old tins and containers for a fresh look just in time for school.

For Every Day

Transform everyday objects into bright and colorful ones. Add reminders to magazines, corral your clutter, and make lunch an event! Turn the little everyday moments into special ones with washi tape.

CRAFT JARS

Keep your ever-expanding craft supplies in check. Use found jars and containers to store beads, glitter, pom-poms, buttons, and more. Jars make for a pretty presentation that you can stack and store easily. Washi tape allows you to label supplies and take note of color names, sources, or other information to keep your clutter to a minimum.

Tape labels work especially well for keeping notes on vintage items that don't have their own packaging. Add notes on where you purchased the item, what the color might be, or even a reminder of what you plan to use it for.

Label infused salts and spices or add dates to those that have a set shelf life so you'll know when to replace expired goods.

SPICE JARS

Add labels to spice jars to keep the clutter in the cupboard to a minimum. Add washi tape labels to jars far in the back so they are instantly recognizable.

STICKY NOTE REMINDERS

Leave notes for loved ones to remind them about plans or just to send them off into the day with a reminder of how much you love them. Add washi tape to ordinary sticky notes for special messages and to-dos.

Create organizational reminders by color-coding sticky notes with a strip of tape.

PLANT MARKERS

Add a little note or reminder to potted plants. Writing on wooden plant markers (or Popsicle sticks) can cause the ink to bleed. Make your message stand out by writing watering instructions, plant names, or sentiments on washi tape with a permanent pen. Add your message to a wooden craft stick, and then nestle it into your plant.

Add washi tape to a wooden ruler or larger stake to mark a growth chart for your plant. This is especially fun for kids as they learn about growing!

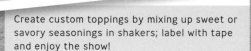

Create custom toppings by mixing up sweet or savory seasonings in shakers; label with tape and enjoy the show!

MOVIE NIGHT

Perfect your popcorn presentation with a colorful border. Use washi tape to change the look of solid dishware and add a punch of pattern or pop of color to your next living room dinner.

Look for tape with days of the week printed on it to create a variation of this calendar. Transform an empty notebook into a planner with days-of-the-week tape.

Add an instant pop of color to plain, plastic hangers, too.

CALENDAR

Using a whiteboard or chalkboard, create a weekly (or monthly) calendar that is cute enough to display! Use washi tape to grid out your calendar, and then write in your days and appointments. Color-code your days, weeks, or months or create something that complements your décor!

HANGERS

Create a keepsake hanger for a special garment, a birthday dress, or a costume. Cover a wooden or plastic hanger with tape strips, overlapping them and trimming along curves and tight corners. Burnish out any ripples or bumps with your fingers as you go. Miss a spot? Just add some more tape!

CLOTHESPINS

Turn this utilitarian object into a reason to add pattern and color to everyday décor. Clothespins can hold a stack of loose receipts or notes, display a photo or small memento, or keep items in place. Why not let function meet design?

Decorated clothespin stands work nicely for an escort card display or table number.

WELCOME MAT

As a sweet surprise for guests and friends, make a temporary welcome mat to greet anyone who enters your home. Spell out a special message for any occasion. Add to the doorway of a child's room or add intention to any spot, like a reading nook or craft space, to make all who enter mindful of where they tread. Change washi tape as often as you need, especially in highly traveled areas.

Use tape to carve out a thinking corner or divide a room for siblings to create individual spaces.

Instead of recipes, write down a memory or intention for the day and label it with the date. File away in a box for a record of the special moments on any given day.

RECIPE CARDS

Using basic index cards, create new traditions by adding washi tape and a recipe to each card. File away in a little box or bundle several to give with a gift of a homemade treat.

Because this tape is transparent, you can also use it to highlight a favorite passage or line. In most cases, it will not be removable on book paper, so test the pages before placing tape.

BOOKPLATES

The best presents given and received are books. Treasure your tomes and label them for your children and their children after them. Place washi tape directly on the front page or add it to store-bought office labels. Create a set to give as a gift or label a book before giving it to a friend.

QUICK-REFERENCE RULER

I can never find my ruler when I need it. I love this preprinted ruler washi tape. Add to a desk, counter, or any surface for quick reference.

Add to the wall to make a height chart that won't permanently mark your space.

Seal a gift box of homemade cookies with washi tape for a sturdy and attractive package.

TAKE IT TO GO

Take-out containers are a great way to box up baked goods or send home leftovers with dinner guests. Package up plain white boxes with pieces of washi tape for an extra-special treat.

PICNIC

The perfect picnic doesn't need to take a lot of planning. A few paper plates and a set of paper napkins can look like fancy china when you add a few stripes of washi tape to your cutlery. Using colors from the paper plates, I added stripes in blue and red to the knives and forks. For a little interest, I added red, blue, and pink stripes to the spoons.

SILVERWARE SAVVY

Wrap sets of cutlery in their napkins so guests can grab them and go. Washi tape is easy to remove on cloth napkins and can be torn open for paper ones. Write names on the tape to create wraps that function as place cards too.

Coordinate elements and unify them with washi tape. If you don't have matching elements, add the same type of tape to everything to bring the look together.

MAGAZINE PAGE MARKERS

Remember what you want to make, cook, buy, read, or do by marking it with a washi tape tag. Call out good ideas, wants and wishes, or must-haves easily. Coordinate colors and add notes to the tabs with a permanent pen.

Use tape tabs instead of folding down page corners you won't ever come back to.

In the Office

Add pattern to sticky notes, mark your cables and cords for easy identification, and brighten your desk with coordinated accessories. Use washi tape to add color to your space in no time.

PHONE CASE

Switch up your phone case whenever the desire for change strikes you. Because this tape is removable, it's the perfect medium for creating endless patterns, images, or even messages. Simply create a design on the back of your phone, trim any excess tape, and then protect it with a clear case!

Create a larger pattern for your laptop computer, and then protect it with a clear case.

IT'S MINE!

No one can argue with a neon pink charger or black-and-white-striped cable. With all the electronic accessories in any given office or home, it's hard to keep track of whose charger is where. Mark your cables and cords with tape for easy and stylish identification. Use washi tape to mark headphones too!

Cover a craft table with washi tape and let kids draw their own design on the layers of tape with a permanent pen.

COFFEE TABLE

Update your look by adding an asymmetrical design to your coffee table. Use wide and regular-width tape to create a dynamic design. Layer solid colors with patterns in a limited palette. Add a pop of color to the legs as well. This is a great way to add color or even cover up a scratch or stain when upcycling a vintage piece of furniture.

DESK ENVY

Who says clutter can't be corralled?
Turn plastic desk organizers and trays
into functional and pretty containers.
Experiment with various patterns
and colors of tape for different looks
and moods.

Use washi tape to decorate magazine holders or label file folders.

OFFICE ACCESSORIES

Black is boring. Even your office supplies can become cute with washi tape! Add the tape to binder clips for a pretty presentation that is both functional and unexpected.

Use solid washi tape and label packets of papers by date, project, or theme.

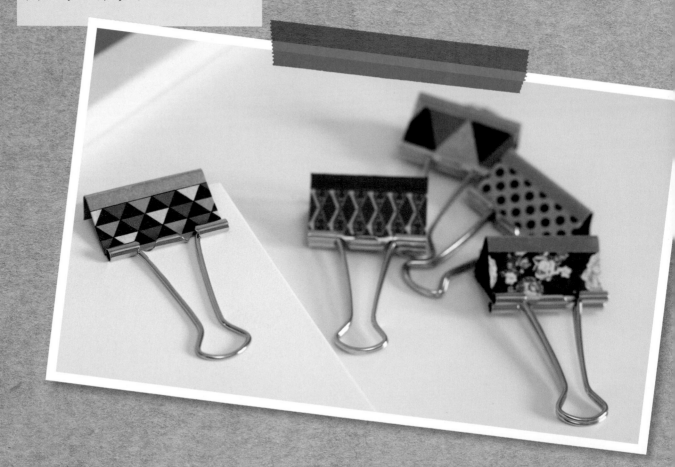

Seasonal

Transform your home into a spooky forest or winter wonderland in a weekend. Add falling snowflakes to your windows and walls, wrap packages in moments, or add a dense forest of trees to your kitchen. Use washi tape to create seasonal décor all year long.

BLOOMING BRANCH

Bring springtime indoors all year round. Add washi tape blossoms in varying shades of pink and blush to bare branches. Place a torn strip of tape on the branch so the branch lands in the middle of the tape. Pinch the tape, crossing the ends at an angle, making a V shape. Twist the tape at the base near the branch to create a soft bloom.

MAKING THE BLOSSOM

Create a miniature version to place on a cake or hang in a window. Add branches down the center of a table for a baby shower, holiday luncheon, tea party, or other springtime gathering. Hang in a window or place upright in a vase for a little added color and reminder of spring.

Make treat baskets for any occasion by changing the look and design with varying types of washi tape.

SOMEBUNNY LOVES YOU

Blank kraft-paper forms are easily found at your local craft store. Transforming these blanks is easy with washi tape. Create a sweet little spring basket for holding tiny toys and treats. Create stripes and patterns by overlapping various styles and sizes of tape for a playful design.

GEOMETRIC WOODEN ORNAMENTS

Using coffee stirrers, create geometric ornaments to hang anywhere. Hold intersections of stirrers together by wrapping a thin strip of washi tape around the joint several times. Add color and pattern by placing regular-width tape along the stirrer. Place ornaments on a tree, hang in a window, or tape to the wall in groups for a striking composition. Reds and golds are perfect for the holidays. Use neons and black for year-round décor that is simple and bold.

Attach sticks in the middle to create star- and snowflake-like shapes to hang.

FLEETING FOREST

Perfect for a hallway or tight corner, make a spooky forest for fall. Use the tape right off the roll and place on the wall to create tall, bare branches. Use varying shades of gray or black for an extra-dark touch. Too scary? Use brown and gold to create a grove of trees for autumn. Add leaves. Create in a child's room for a cozy woodland corner.

Create a forest of trees on a window to complete the look.

SCRAPPY WREATH

Put all those tape trimmings to good use. Layer scraps and torn pieces of tape on a piece of paper or on a card to create a small and scrappy wreath. Hang it on the door or make a miniature for a seasonal card. It's great to do with kids, but fun for adults, too!

Use your tape scraps as the foundation for a piece of art. Place tape trimmings on blank pages to create a springboard for inspiration.

WINE WRAP

Forget the standard, clumsy wine wrap or those tacky organza bags. With a strip of any solid wrap (I love kraft paper), create a sleeve that extends just beyond the wine label. Add several strips of washi tape to hold it in place. Layer the tape to create a design that finishes off your wrapping. Use tape with a preprinted message or graphic for a personalized touch.

Mark the base of your wineglass with a strip of tape as a wine marker at your next party.

ADVENT CALENDAR

The countdown to Christmas is one of the most thrilling times of year. Make the wait extra special with an Advent calendar. Instead of opening a store-bought calendar to reveal yet another piece of mock chocolate, create your own pockets to hold treasures. Decorate envelopes, muslin bags, or small lunch bags with washi tape. Add candies, notes, and other small gifts to each pouch. Hang along a length of twine for a simple presentation.

You can use washi tape with numbers to create your countdown or use rubber stamps to mark the days until Christmas.

Use this idea to create count-downs for birthdays, anniversaries, or vacations.

SNOWFLAKES ON WINDOWS

It never snows where I live, but I can create my own winter wonderland with washi tape. Prepare snowflakes and save them on waxed paper or create them right on the window or wall. Use wintry greens and blues, shimmering silvers, or gray-and-white-patterned tape for a soft feel. Vary thick and thin tapes for hearty snowflakes as well as wispy, delicate ones.

Create stars or constellations and place on a ceiling to create a summer-night sky indoors instead.

TREE ON WALL

For tiny spaces, a real tree for Christmas isn't always possible. But the experience of finding presents under the tree should never be missed. Make your own tree without taking up any extra space. Use washi tape to make a tree silhouette and place small gifts underneath.

A washi tape tree could also be used as foundation for an Advent calendar.

Resources

MY WEBSITE!

www.ccerruti.com

To see more of my work, follow me on Instagram @ccerruti and get inspired / by my pins!

I would love to see what you make with washi tape. Post to Instagram and hashtag #101washitape, @ccerruti. http://pinterest.com/ccerruti/

WISHY WASHI TAPE

www.wishywashi.com

This is a beautifully curated site with a huge selection and variety of washi tape. They also sell the larger home décor washi (MT Casa) rolls at their sister site, www.mtcasa.com. Check this site frequently because they're always adding new patterns and styles.

MAIDO STATIONERY

www.mymaido.com

This is a great Japanese stationery store with paper goodies and washi tape galore. This is the first place I saw washi tape, and it is still my favorite place to see what's new and browse. Unfortunately, they don't sell washi tape on their website, but they have locations in San Francisco, San Jose, Seattle, Los Angeles, and New York.

ONCE AROUND

www.oncearound.com

Mill Valley, California

This is one of my favorite local craft stores. This beautiful shop has a little bit of everything, including a whole bookcase of washi tape. In addition to their wonderful selection, they break up sets of washi tape so you can buy rolls individually. Love it!

CU.TE. TA.PE

www.cutetape.com

Cute Tape stocks washi tape, but they also carry a lot of great blanks for decorating, like mailing tags, coasters, clothespins, favor bags, envelopes, and cards.

LIFESTYLE CRAFTS

www.lifestylecrafts.com

Lifestyle Crafts carries sheets of washi tape called shape n tape that you use to create washi tape stickers. Use these sheets with craft punches or cut custom shapes with a die cut machine or scissors.

Places like Paper Source, Michaels, and other craft stores are carrying washi tape more and more. Often they stock brands of tape that will be described as decorative tape or paper tape but have the properties of washi tape. Don't be afraid to try them out and see how they work for your project.

Paper Source

www.paper-source.com

Michaels

www.michaels.com

Acknowledgments

Thank you, Mary Ann, for choosing me to write this book. Best Valentine's Day EVER.

Thank you, Jonathan, for being as excited as I am about washi tape, for sending me pictures and ideas throughout the process, and for being open and thoughtful along the way. Thank you for bringing washi tape to meetings and converting the Quarry office into washi tape users! We will conquer the world with washi tape!

Thank you, friends and family, for being enthusiastic about this book project, for shouting out ideas midconversation, and for tolerating my undying enthusiasm for washi tape.

Thank you, Mom and Dad, for always supporting me, wherever my creative (and wandering) pursuits lead me.

Thank you, Matt, for working with me on this book and for taking such awesome photos. Thank you for talking me through the late nights of washi tape madness and for encouraging me every step of the way.

Thank you, Dickson and Katina, for your sun-filled apartment, which felt like a magical tree house instead of a shoot location.

Thank you, Ava, for providing the perfect backdrop for many of these projects; your space is lovely!

Thank you, Celia, Joe, and Cheech . . . you are each truly a peach!